FOREST CREATURES

WILD BOARS IN THE FOREST

DONNA REYNOLDS

Gareth Stevens PUBLISHING

Please visit our website, www.garethstevens.com. For a free color catalog of all our high-quality books, call toll free 1-800-542-2595 or fax 1-877-542-2596.

Library of Congress Cataloging-in-Publication Data
Names: Reynolds, Donna, author.
Title: Wild boars in the forest / by Donna Reynolds.
Description: New York, NY : Gareth Stevens Publishing, [2023] | Series: Forest creatures | Includes index.
Identifiers: LCCN 2022008205 | ISBN 9781538279298 (library binding) | ISBN 9781538279274 (paperback) | ISBN 9781538279281 (set) | ISBN 9781538279304 (ebook)
Subjects: LCSH: Wild boar–Juvenile literature.
Classification: LCC QL737.U58 R49 2023 | DDC 599.63/32–dc23/eng/20220310
LC record available at https://lccn.loc.gov/2022008205

Published in 2023 by
Gareth Stevens Publishing
29 East 21st Street
New York, NY 10010

Portions of this work were originally authored by Harper Avett and published as *Wild Boars*. All new material in this edition was authored by Donna Reynolds.

Editor: Jennifer Lombardo
Designer: Rachel Rising

Photo credits: cover, p. 9 Martin Prochazkacz/Shutterstock.com; cover, pp. 1, 3, 4, 6, 8, 10, 12, 14–16, 18, 20, 22–24 lavendertime/Shutterstock.com; cover, p. 1 furtseff/Shutterstock.com; cover, pp. 1, 3, 4, 6, 8, 10, 12, 16, 18, 20, 22–24 Zvezdesign/Shutterstock.com; p. 5 (top) Photosite/Shutterstock.com; p. 5 (bottom) Dmitry Kalinovsky/Shutterstock.com; p. 7 sma1050/Shutterstock.com; p. 11 taviphoto/Shutterstock.com; p. 13 Ben Schonewille/Shutterstock.com; p. 14 Eric Isselee/Shutterstock.com; p. 15 MM.Wildlifephotos/Shutterstock.com; p. 17 Bildagentur Zoonar GmbH/Shutterstock.com; p. 19 (main photo) Michal Kimmel/Shutterstock.com; p. 19 (top inset) Goran Jakus/Shutterstock.com; p. 19 (bottom inset) Photoagriculture/Shutterstock.com; p. 21 eAlisa/Shutterstock.com.

Printed in the United States of America

CPSIA compliance information: Batch #CSGS23: For further information, contact Gareth Stevens, New York, New York, at 1-800-542-2595.

CONTENTS

Words in the glossary appear in **bold** type the first time they are used in the text.

BOARS AND PIGS

Wild boars were some of the first animals early humans **domesticated**, turning them into the pigs you can see on farms today. The biggest difference is in the way they look. Domesticated pigs are best known for having mostly pink skin and very little hair.

Wild boars have a lot more hair than almost all domesticated pigs. They also have longer tails, which aren't curly like most domesticated pigs' tails. Their heads and legs are longer too.

FIND OUT MORE!

Farmers use the word "boar" to mean a young male domesticated pig, but all wild pigs are called boars, no matter what age or sex they are.

The Mangalica pig is a **breed** of pig that's almost as woolly as a sheep! Other domesticated pigs have much less hair.

TRAVELING BOARS

Wild boars can be found in North and South America as well as Australia, but they aren't **native** to those places. They're native only to Europe, Asia, and some parts of North Africa. They were introduced, or brought, to other places by humans.

In the 1500s, the Spanish brought wild boars to their American colonies. The boars were meant to be eaten. Many escaped from their pens, so the **species** spread to new places. Today, wild boars have been found in at least 45 U.S. states.

FIND OUT MORE!

The Spanish explorer Hernando de Soto was the first person to bring wild boars to what is now the southeastern United States.

Shown here is a wild boar in Malaysia, one of its native habitats.

LIFE IN THE FOREST

A forest is a great place for a wild boar to live because it provides shelter and food. Roots and leaves are some of the wild boar's favorite foods, and the forest is full of them.

However, boars aren't picky about where they live. Aside from forests, they're known to live in grasslands such as savannas. They also live in wetlands such as marshes. Wild boars live in the Everglades, which is a large area of wetlands in Florida.

FIND OUT MORE!

If a domesticated pig gets out into the wild, it will soon start to look and act more like a wild boar! Its hair will grow longer, and it'll get more **aggressive**.

Given the choice, a wild boar would rather live in the forest than out in the open.

WHAT DOES A BOAR LOOK LIKE?

Wild boars can grow to more than 7 feet (2.1 m) long. On average, they weigh about 180 pounds (82 kg). They have black, brownish, or white hair. The hair is thick and rough. Some boars have patterns in their coat, and most have longer hairs down the center of their back.

Many wild boars have two long teeth that are called tusks. Their tusks can grow as long as 9 inches (23 cm). Both male and female boars can have tusks, but they're shaped differently.

FIND OUT MORE!

The largest wild boar on record was shot by an 11-year-old boy in 2007. It was more than 9 feet (2.7 m) long and weighed 1,051 pounds (477 kg)!

The tusks of male boars *(left)* are much more noticeable than those of the females *(right)*.

WILD BOAR FAMILIES

Most animals mate, or come together to make babies, only once a year. However, boars in warm places mate at any time of the year. A mother boar may have 4 to 12 babies, or piglets, at a time. She may give birth more than two times a year.

A mom and her piglets make up one family. Multiple families often live together in huge groups called sounders. Wild boar sounders can have up to 100 members. Sounders are matriarchal, meaning that one older female boar is the leader of the group.

FIND OUT MORE!

Adult male wild boars are **solitary** except when they're mating. Male piglets stay in a sounder until they're about three years old, then go off on their own.

Wild boar piglets are born with stripes, which help them stay hidden in the forest.

NOT PICKY EATERS

Part of the reason why wild boars can live just about anywhere is because they'll eat just about anything. They're omnivorous, which means they eat both plants and meat.

Wild boars in warmer places are most active at night. They **forage** for roots, nuts, fruit, and other plant parts, which make up 90 percent of their diet. They also eat bugs, worms, and small animals such as birds, mice, and snakes. They're even known to eat carrion, or the rotting bodies of dead animals.

AN OMNIVOROUS DIET

PLANTS	ANIMALS
acorns	birds
corn	earthworms
mushrooms	eggs
oats	goats (young or sick)
watermelon	rabbits

Above are just a few things wild boars are known to eat.

USING THEIR NOSE

Wild boars can't see very well. To make up for this, they have an amazing sense of smell. This helps them forage for food. It also helps them know when there are predators close by.

Boars can smell things as deep as 25 feet (7.6 m) underground. They can also smell things 5 to 7 miles (8 to 11.3 km) away. They can dig down into the ground 3 feet (0.9 m) deep. This is called "rooting" because they're often looking for roots to eat.

FIND OUT MORE!

Wild boars are hard to hunt because as well as their great noses, they have great brains. They're smarter than dogs, and they quickly learn human hunters' tricks.

Boars keep their nose close to the ground when they're foraging.

A PEST TO FARMERS

Farmers see wild boars as **pests** because these animals can be very destructive, or harmful. When they root into the ground, it erodes, or wears away at, soil. This makes it harder to grow crops.

Wild boars will also eat crops from farms. They can clear fields of crops by turning up the soil and eating along the way. They eat seeds and young plants. This costs farmers a lot of money because they have less food to sell.

FIND OUT MORE!

Wild boars carry diseases that they can pass on to domesticated pigs on farms. They can also carry **rabies**, which is dangerous, or unsafe, for humans and animals.

Wild boars cause a lot of problems on farms.

BOARS AND PEOPLE

Because wild boars aren't native to the United States, they have serious effects on the **ecosystem**. That's because native animals have to **compete** with them for food and water. For this reason, some states, such as Texas, let people hunt as many boars as they want.

Boars aren't usually aggressive toward humans. They can smell people coming in the woods a long way away, so they have a lot of time to hide. However, if a boar is hurt or scared for its piglets, it may attack.

FIND OUT MORE!

Not many animals will attack a wild boar. Hunting by humans helps keep their numbers down so there aren't too many of them.

Some people use dogs to hunt wild boars, but this can be dangerous for the dogs.

GLOSSARY

aggressive: Showing a readiness to attack.

breed: A kind of animal or plant that is found only under human care and is different from related kinds.

compete: To try to win something that someone else is also trying to win.

domesticated: Bred and raised for use by people.

ecosystem: A natural community of living and nonliving things.

forage: To search for food to eat.

native: Living or growing naturally in a particular place.

pest: A plant or animal that is harmful to humans or property.

rabies: A sometimes deadly disease that affects the central nervous system. It's carried in the spit of some animals.

solitary: Living alone.

species: A group of plants or animals that are all of the same kind.

FOR MORE INFORMATION

BOOKS

Leighton, Christina. *Pigs*. Minneapolis, MN: Bellwether Media, 2018.

Rheinstrom, David. *Animal Attack! A Wildlife Survival Guide*. New York, NY: Tangerine Press, 2019.

WEBSITES

DK Find Out! Pigs
www.dkfindout.com/us/animals-and-nature/pigs
This website includes a photo of a wild boar with labeled body parts.

Kiddle: Wild Boar Facts for Kids
kids.kiddle.co/Wild_boar
Learn more about boars and see pictures from multiple points in history.

Publisher's note to educators and parents: Our editors have carefully reviewed these websites to ensure that they are suitable for students. Many websites change frequently, however, and we cannot guarantee that a site's future contents will continue to meet our high standards of quality and educational value. Be advised that students should be closely supervised whenever they access the internet.

INDEX